statistics
made easy

Alan Graham

Hodder Education
338 Euston Road, London NW1 3BH.

Hodder Education is an Hachette UK company

First published in UK 2011 by Hodder Education.

British Library Cataloguing in Publication Data: a catalogue record for this title
is available from the British Library.

10 9 8 7 6 5 4 3 2 1

The publisher has used its best endeavours to ensure that any website
addresses referred to in this book are correct and active at the time of going
to press. However, the publisher and the author have no responsibility for the
websites and can make no guarantee that a site will remain live or that the
content will remain relevant, decent or appropriate.

The publisher has made every effort to mark as such all words which it
believes to be trademarks. The publisher should also like to make it clear that
the presence of a word in the book, whether marked or unmarked, in no way
affects its legal status as a trademark.

Every reasonable effort has been made by the publisher to trace the copyright
holders of material in this book. Any errors or omissions should be notified in
writing to the publisher, who will endeavour to rectify the situation for any
reprints and future editions.

Hachette UK's policy is to use papers that are natural, renewable and
recyclable products and made from wood grown in sustainable forests.
The logging and manufacturing processes are expected to conform to the
environmental regulations of the country of origin.

www.hoddereducation.co.uk

Typeset by MPS Limited, a Macmillan Company.
Printed in Great Britain by CPI Cox & Wyman, Reading.

Contents

1   Some basic maths   2

2   Graphing data   12

3   Summarizing data   22

4   Choosing a sample   32

5   Collecting information   40

6   Spreadsheets to the rescue   48

7   Regression: describing relationships between things   56

8   Correlation: measuring the strength of a relationship   64

9   Chance and probability   74

10   Deciding on differences   84

1

some basic maths

If you flick through the rest of this book, you will find that complex mathematical formulas and notations have been kept to a minimum. However, it is hard to avoid such things entirely in a book on statistics, and if your maths is a bit rusty you should find it worth spending some time working through this chapter. The topics covered are algebra, coordinates and statistical notation and they will offer a helpful mathematical foundation, particularly for Chapters 3, 7 and 8.

Algebraic symbols and notation crop up quite a lot in statistics. For example, in Chapter 7 when you come to find the 'best fit' line through a set of points on a graph, you will learn how to express this line algebraically, as an equation.

In this chapter, you will look briefly at two important symbols in statistics, Σ (sigma) meaning 'the sum of' and $\bar{x}$ (x-bar) which is the mean.

Algebra

Can you remember how to convert temperatures from degrees Celsius (formerly known as Centigrade) to degrees Fahrenheit? If you were able to remember, the chances are that your explanation would go something like this:

> *Suppose the original temperature was, say, 20°C. First you multiply this by 1.8, giving 36 and then you add 32, so the answer is 68.*

The explanation above, which is perfectly valid, is based on the common principle of using a *particular* number, in this case a temperature of 20°, to illustrate the *general* method of doing the calculation. It isn't difficult to alter the instructions in order to do the calculation for a different number, say 25 – simply replace the 20 with 25 and continue as before. What we really have here is a formula, in words, connecting Fahrenheit and Celsius. A neater way of expressing it would be to use letters instead of words, as follows:

$$F = 1.8C + 32$$

(where F = temperature in degrees Fahrenheit and C = temperature in degrees Celsius).

Reading the formula aloud from left to right, you would say 'F is equal to one point eight C plus thirty-two'.

The formula is equivalent to the word description shown earlier but here the letter C stands for whatever number you wish to convert from °C to °F. The F is the answer you get expressed in °F. The choice of letters for a formula is quite arbitrary – X and Y are particular favourites – but it makes sense to choose letters so that it is easy to remember what they stand for (hence F for Fahrenheit and C for Celsius in the formula shown above). The main point to be made here is that, in algebra, each letter is a sort of place-holder for whatever number you may wish to replace it with. A formula such as the one above provides a neat summary of the relationship that you are interested in and shows the main features at a glance.

There are certain conventional rules in algebra which you need to be clear about. Firstly, have a look at the following examples where a number and a letter are written close together:

$$5x; 1.8\,C; -3y$$

Although it doesn't actually say so, in each case the two things, the number and the letter, are to be *multiplied* together. Thus, $5x$ really means 5 times x. Similarly, $1.8\,C$ means 1.8 times C, and so on.

A formula is really another word for an *equation*. Usually a formula is written so that it has a single letter on the left of the equals sign (in this case the F) and an expression containing various numbers and letters on the right. We say that the equation $F = 1.8\,C + 32$ is a *formula for F* (F is the letter on the left of the equals sign) *in terms of C* (C is the only letter, in this case, on the right-hand side of the equals sign).

To *satisfy* a formula or an equation is to find values to replace the letters which will make the two sides of the equation equal. For example, the values $C = 0$, $F = 32$ will satisfy the equation above. We can check this by *substituting* the two values into the equation, thus:

$$F = 1.8\,C + 32$$
$$32 = 1.8 \times 0 + 32 \dots \text{which is true!}$$

In the previous example, notice that the temperature in °C is multiplied by 1.8 before you add the 32. In general, where you appear to have a choice of whether to add, subtract, multiply or divide first, the multiplication and division take precedence over addition and subtraction. This would be true even if the formula had been written the other way round, like this:

$$F = 32 + 1.8\,C$$

This may seem odd but the reason for the convention is simply that, otherwise, the formula would be ambiguous. So it doesn't matter which way round the formula is written; provided the multiplication is done first, the result is still the same.

You might be wondering what happens if you want a formula involving multiplication and addition but where you wish the addition to be done before the multiplication. This is where the use of brackets comes into algebra. An example of this is where the temperature conversion formula is rearranged so that it will convert from degrees Fahrenheit into degrees Celsius. Remember that if you reverse a formula, not only do you have to reverse the operations (× becomes ÷ and + becomes −) but you must also reverse the sequence of the operations.

There are certain problems if we write the new conversion formula as:

$$C = F - 32 \div 1.8$$

If this formula were to be entered directly into a calculator with algebraic logic, it would give the 'wrong' answer. The reason is that, because division has a higher level of precedence than subtraction, a calculator with algebraic logic would divide 32 by the 1.8 before doing the subtraction. Putting in brackets, as shown below, takes away the ambiguity and forces the calculator to perform the calculation in the required sequence.

$$C = (F - 32) \div 1.8$$

To end this section, here are a few of the common algebraic definitions that you may need to be familiar with.

* **Expression:** $3X^2 - 4X + 5XY + 3X$ is an example of an expression.
* **Term:** There are four terms in the previous expression. They are, respectively, $3X^2$, $- 4X$, $5XY$ and $3X$.
* **Sign:** The sign of a term is whether it is positive or negative. Only the second of these four terms is negative; the others are positive. When we are writing down a positive term on its own we don't normally bother to write in the '+' sign before it. Thus we would write $5XY$, rather than $+5XY$.
* **Term type:** This refers only to the part of the term that is written in letters. Thus, the first term in the expression

above is an 'X-squared' term, the second is an 'X' term, and so on.

* **Coefficients:** The coefficient of a term is the number at the front of it. Thus, the coefficient of the first X term is -4. The coefficient of the XY term is 5, and so on. The coefficient tells you how many of each term type there are.

* **Like term:** Two terms are said to be 'like terms' when they are of the same term type. Thus, since the second and fourth terms in the expression above are both 'X terms', they are therefore 'like terms'. The phrase 'collecting like terms' describes the process of putting like terms together into a single term. For example the second and fourth terms, $-4X + 3X$, can be put together simply as $-X$. This result arises from adding the two coefficients, -4 and 3, giving a combined coefficient of -1. We don't normally write this as $-1X$, but rather as $-X$.

* **Simplifying expressions:** This is the general term for collecting like terms. It is a simplification in that the total number of terms is reduced to just one each of every term type.

* **Expanding:** This normally refers to expanding out (i.e. multiplying out) brackets. Thus, the expression $5(2X + 3) - 3(6 - X)$ could be expanded out to give $10X + 15 - 18 + 3X$. The purpose of doing this is usually to enable further simplification. Thus, in this case we can collect together the two X terms and the two number terms to give the simplified answer $13X - 3$.

* **Equations:** Equations look rather like expressions except that they include an equals sign, '='. In fact, an equation can be defined as two expressions with an equals sign between them. The expression to the left of the equals sign is, for obvious reasons, called the 'left-hand side', or 'LHS', while the other expression to the right of the equals sign is the 'right-hand side' or 'RHS'. Here is an example of an equation: $2X - 3 = 4 - 6X$. Often, however, equations are written so that the RHS expression is zero. For example, $3 - 4X + X^2 = 0$.

* **Solving equations:** Equations usually exist in mathematics textbooks to be 'solved'. Solving an equation means finding the value or values for X (or whatever the unknown letter happens to be) for which the equation is true. For example, the simple equation $3X - 5 = 7$ holds true for one value for X. The solution to this equation is $X = 4$. This can be checked by 'substituting' the value $X = 4$ back into the original equation to confirm that the LHS equals the RHS. Thus, we get LHS = $3 \times 4 - 5$, which equals 7. This is indeed the value of the RHS. If you were to try any other value for X say $X = 2$, the equation would not be 'satisfied'. Thus, LHS = $3 \times 2 - 5$, which equals 1 not 7. The equation $3 - 4X + X^2 = 0$ holds true for two values of X, $X = 1$ and $X = 3$, so there are two solutions to this equation.

Graphs

Perhaps more than any other mathematical topic, the basic idea of a graph keeps cropping up in statistics. The reason for this is to do with how the human brain processes information. For most people, scanning a collection of figures, or looking at a page of algebra doesn't reveal very much in the way of patterns. However, our brains seem to be much better equipped at seeing patterns when the information has been re-presented into a picture. The crucial difference is that in a picture (a graph or diagram, for example) each fact is recorded onto the page by its *position*, rather than by writing some peculiar, arbitrary squiggle (a number or a letter).

Basically there are three pieces of information that you need to provide in order to fix exactly where a point is located. These are:
* which (original) point on the page you started measuring from – known as the 'origin';
* how far from the origin the point is *across* the page – the *horizontal* distance; and
* how far from the origin the point is *up* the page – the *vertical* distance.

In mathematics, graphs are organized around these three principles. The 'page' is laid out in a special way so that the starting point (the origin) is clearly identified – normally in the bottom left hand corner. The two directions, horizontal and vertical, are normally drawn out as two straight lines and are known, respectively, as the X axis and the Y axis. Finally, in order to keep things as simple as possible, the two axes are provided with a scale of measure.

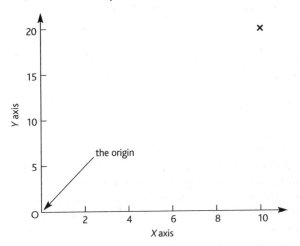

Figure 1.1 *The axes of a graph.*

So, provided the page is laid out with the origin clearly identified, and the axes are suitably scaled, the point marked with an × in Figure 1.1 can be described by giving just two numbers, in this case 10 and 20. These two numbers are called the coordinates of the point and they are usually written inside brackets, separated by a comma, like this:

$$(10, 20)$$

The first number inside the brackets, known as the X coordinate, measures how far you have to go along the X direction from the origin. The second number, the Y coordinate, gives the distance you go up the Y direction.

Finally, have a look at where the $Y = 2X - 3$ line passes through the Y axis. It is significant that it intersects the Y axis at the same value as the number term (-3) in the equation. This is a useful fact to remember about linear equations. The value -3 is known as the 'Y intercept'. The other number in the equation is the one associated with the X, in this case the number 2, and it will tell you how steep the line is – known as the 'slope' or 'gradient' of the line. It is a measure of how far the line goes up in the Y direction for every unit distance it goes across in the X direction. Expressed in more mathematical language, we can summarize all this by the following statement:

The general linear equation, $Y = a + bX$, has a *slope* of b and an *intercept* of a.

Notation

There are one or two special symbols and notations in statistics that are in common use and are well worth familiarizing yourself with. If you have a specialized statistical calculator, you will already have run into some of these. The two described here are Σ and $\overline{X}$.

(Capital) sigma Σ

The first important symbol is the Greek letter capital sigma, written as Σ. In order to keep things interesting for you, statisticians have not one, but two symbols called 'sigma' in regular use. The lower case Greek letter sigma is written as 'σ' and it is normally used in statistics to refer to a measure of spread, called the 'standard deviation' (which is explained in Chapter 3).

This symbol, Σ, is an instruction to add a set of numbers together. So, ΣX means 'add together all the X values'. Similarly, ΣXY means 'add together all the XY products'.

X Bar, $\overline{X}$

The symbol $\overline{X}$ is pronounced 'X bar' and refers to the mean of the X values. A 'mean' is the most common form of average, where you add up all the values and divide by the number of values you had. To take the previous example, the five X values are 0, 1, 2, 3, and 4. In this example, the mean value of X, is written as:

$$\overline{X} = \frac{\Sigma X}{n}$$

Thus, $\overline{X} = \dfrac{0 + 1 + 2 + 3 + 4}{5} = \dfrac{10}{5} = 2$

2

graphing
data

Increasingly these days, graphs, charts and plots are created on a computer rather than by using pencil and paper. There is no doubt that, using a computer, you can achieve a most attractive chart quickly and easily. However there is a downside. The only contribution offered by the machine is the technical one of transforming the data into the chart of your choice. What it cannot do is to decide whether one particular chart offers a better way to represent your information than another, or indeed, whether your dataset is suitable for presenting visually, at all.

There are two key questions you need to ask when making a choice of graph or chart:
* Does it work for the type of data you are depicting?
* Does it meet your purpose in creating it – in particular, is your purpose to *summarize, make a comparison* or *explore an inter-relationship*?

This chapter introduces the main features of the following graphs – bar charts, pie charts, histograms, stemplots, scattergraphs and time graphs. Boxplots are explained in Chapter 3.

Bar charts

The bar chart is probably the most familiar of all the graphs that you are likely to see in newspapers or magazines. Figure 2.1 is a typical example, which summarizes the popularity of different methods of contraception as reported to a particular family planning clinic.

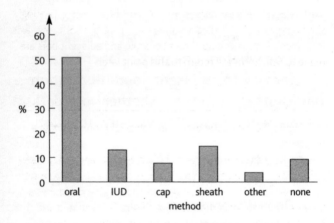

Figure 2.1 *Family planning clinic: methods of contraception recommended or chosen.*
Source: estimated from UK national data.

The chief virtue of a bar chart is that it can provide you with an instant picture of the relative sizes of the different categories. This enables you to see some of the main features from a glance at the bar chart – for example that oral contraceptives are by far the most popular method, that roughly twice as many use the sheath as the cap and so on. (Incidentally, since these figures have been collected by a family planning clinic, they reflect only the preferences of those people who attended and it would be unwise to infer that they are typical of all family planning clinics or of the population as a whole.) There are a number of characteristics of a bar chart that are worth drawing attention to.

Firstly, you can see that the bars have all been drawn with the *same width*. This makes for a fair comparison between the bars in that the height of a particular bar is a measure of the frequency with which that category occurs.

Next, notice that it has been drawn with *gaps* between the adjacent bars. The reason for this is to emphasize the fact that the different methods of contraception listed along the horizontal axis are quite *separate categories*. This contrasts with a seemingly similar type of graph called a histogram, where the horizontal axis is marked with a continuous number scale and adjacent bars are made to touch. We will return to this point later.

Pie charts

Figure 2.2 shows the data from Table 2.1 represented in a pie chart.

As you can see from Figure 2.2, a pie chart represents the component parts that go to make up the complete 'pie' as a set of different-sized slices. Drawing a pie chart by hand requires a little care, both in how the size of each slice is calculated and how it is drawn. The size of each slice as it appears on the pie chart will be determined by the angle it makes at the centre of the pie. You may remember that a full turn is 360° (360 degrees), so the angles at the centre in all the slices must add up to 360°. Taking Figure 2.2 as an example, the IUD slice should take up 13% of the pie, so the angle at the centre of that particular slice must be 13% of the entire 360°. If you have a calculator to hand calculate:

$$360 \times \frac{13}{100} \text{ to get the answer } 46.8°$$

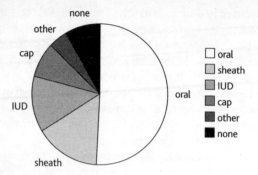

Figure 2.2 *Pie chart showing methods of contraception recommended or chosen. Source: Table 2.1.*

Notice that the legend box has been included on the right of the pie chart in Figure 2.2. It isn't strictly necessary here, since the categories have also been written beside the corresponding slices of the pie. Normally either one or other of these approaches would be used as a means of identifying each slice of the pie.

Histograms

The histogram in Figure 2.3 shows the proportions of babies born to mothers at different ages, in a typical UK hospital.

A key distinction between the histogram in Figure 2.3 and the bar chart described earlier is that here the adjacent bars are made to touch. The reason for this is to emphasize the fact that the horizontal axis on a histogram represents a continuous number scale. It is important to be aware that the bars on a histogram do not represent separate categories, as they do on a bar chart, but, rather, adjacent intervals on a number line. In other words, although superficially the bar chart and the histogram look similar, they are designed to represent two quite different types of data. The bar chart is useful for depicting separate *categories*, while the histogram describes the 'shape' of data that have been *measured on a continuous number scale*. The distinction between these two types of data is an important one and will be returned to in the next

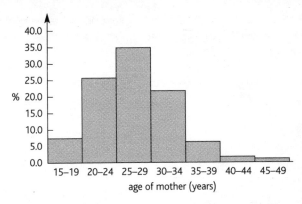

Figure 2.3 *Histogram showing the number of live births by age of mother.*
Source: estimated from national data.

chapter. In the example above, the data have been grouped into intervals of five years on the horizontal axis. This is known as the class interval and in Figure 2.3 the class intervals are all equal.

Finally, let's turn our attention to the scale on the vertical axis which shows the percentage of women contained in each age interval. Where all the class intervals are the same width, as is the case in Figure 2.3, it seems to be reasonably clear what these percentage figures refer to. However, where the class intervals are unequal, a number scale on the vertical axis is less meaningful since the height of each bar must also be interpreted in terms of how wide it is. The key property which can be relied on is the area of each bar since, the total area of a histogram must be preserved no matter how the class intervals are altered.

Stemplots

Stemplots, sometimes called 'stem-and-leaf' diagrams, can often be used as an alternative to histograms for representing numerical data. Table 2.1, for example, gives some numerical data printed in a national newspaper over a period of several consecutive days.

They have been taken from a Birthday column and reveal the ages of 40 women deemed to be sufficiently famous to warrant inclusion in the newspaper.

Table 2.1 *The ages of 40 'famous' women listed in the Birthday column of a national newspaper*

77	32	55	55	59	67	55	60	51	82
66	66	100	29	61	47	52	46	53	63
74	47	58	72	55	50	36	52	58	48
80	41	54	53	70	68	42	62	98	45

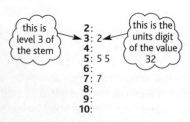

Figure 2.4 *Incomplete stemplot representing the first four items from Table 2.1.*

It makes sense to sort the stemplot so that the 'leaves' on each level are ranked from smallest to largest, reading from left to right (Figure 2.4). The final version, called a 'sorted' stemplot, is shown in Figure 2.5.

```
 2: 9
 3: 26
 4: 1256778
 5: 01223345555889
 6: 01236678
 7: 0247
 8: 02
 9: 8
10: 0
```

10: 0 is 100 years

Figure 2.5 *'Sorted' stemplot based on the data from Table 2.1.*

As you can see, the stemplot looks a bit like a histogram on its side. However, it has certain advantages over the histogram, the main one being that the actual values of the raw data from which it has been drawn have been preserved. Not only that but, when the stemplot is drawn in its sorted form, the data are displayed in rank order.

Scattergraphs

Scattergraphs (sometimes known as scatterplots or scatter diagrams) are useful for representing paired data in such a way that you can more easily investigate a possible relationship between the two things being measured. Paired data occur when two distinct characteristics are measured from each item in a sample of items. Table 2.2, for example, contains paired data relating to twelve countries of the European Union (EU) – the area (in 1000 km^2) of each country and also their estimated populations (in thousands) for the year 2007. Figure 2.6 shows the same data plotted as a scattergraph.

Table 2.2 *Area and estimated population of 12 EU countries*

Country	Area (1000 km²)	Estimated population for 2007 (million)
Belgium	30.5	10.4
Denmark	43.1	5.5
France	544	63.7
Germany	357	82.4
Greece	132	10.7
Ireland	68.9	4.1
Italy	301.3	58.1
Luxembourg	2.6	0.5
Netherlands	41.2	16.6
Portugal	92.1	10.6
Spain	504.8	40.4
United Kingdom	244.1	60.8

Scattergraphs are a particularly important form of representation in statistics as they provide a powerful visual impression of what sort of relationship exists between the two things under consideration. For example, in this instance, there seems to be a clear pattern in the distribution of the points, which tends to confirm what you might have already expected from common sense, namely that countries with large areas tend also to have large populations.

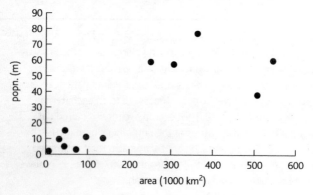

Figure 2.6 *Scattergraph showing the relationship between population and area in 12 EU countries.*

Time graphs

Figure 2.7 shows typical time graphs. Time is almost always measured along the horizontal axis and in this case the vertical axis shows the number of men and women (million) unemployed between 1971 and 2006.

There is an important difference between time graphs and scattergraphs. With time graphs, each point is recorded in regularly spaced intervals going from left to right along the time axis. Since they therefore form a definite sequence of points, it makes sense to join them together with a line, as has been done here. However, points plotted on a scattergraph are not consecutive and it would be incorrect to attempt to join the points on a scattergraph in this way.

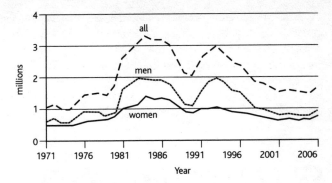

Figure 2.7 *Unemployment by sex in the UK, 1971–2006.*

3

summarizing
data

We live in a world of rapidly growing collections of data. Information is being amassed and communicated on an ever-wider range of human activities and interests. Even if supported by a computer database, a calculator or a statistical package, it is hard for the 'average person' to gain a clear sense of what this information might be telling them. A crucial human skill is to be selective about the data that we choose to analyse and, where possible, to summarize the information as briefly and usefully as possible. In practice, the two most useful questions that will help you to summarize a mass of figures are:

* What is a typical, or average value?
* How widely spread are the figures?

This chapter looks at a few of the most useful summary measures under these two headings, *average* and *spread*.

Introducing averages – finding a middle value

The next section looks at three particular averages and how they are calculated.

Mode

The mode of a batch of data is usually defined as the most frequently occurring item. The procedure for finding the mode is as follows:

i identify all the distinct values or categories in the batch of data;

ii make a tally of the number of occurrences of each value or category;

iii the mode is the value or category with the greatest frequency.

A simple mean ($\overline{X}$)

The mean, or the arithmetic mean, as it is sometimes called, is the best known average which can be defined as 'the sum of the values divided by the number of values'. The mathematical symbols used to describe the mean were explained in Chapter 1. Here they are again.

the symbol for the mean, pronounced 'X bar'

$$\overline{X} = \frac{\sum X}{n}$$

the Greek letter capital sigma, meaning 'the sum of'

Median (MD)

The median of a set of, say, nine values is found by ranking the values in order of size and choosing the middle one – in the case of nine values it is the value of the fifth one. Note that a common mistake is to say that the value of the median of nine items is 5. It needs to be stressed that the median is *the value* of item number 5 when all nine items are ranked in order and is not its rank number. With an odd number of items, like 11 or 37 or 135, there will always be a unique item in the middle (respectively the 6th, 19th and 68th). A simple way of working this out is to add one to the batch size and divide by 2. For example, for a batch of 37 items, $(37 + 1)/2 = 19$, so the median is the value of the 19th item. However, where the batch size of the data is even, there is not a unique item in the middle. In such circumstances, the usual approach is to choose the two middle values and find their mean.

Example

A count of the contents of eight boxes of matches produced the following results:

  Number of matches   52 49 47 55 54 51 50 50

Find the median number of matches in a box.

Sorting the numbers in order of size produces the following:

47 49 50 50 51 52 54 55

the two middle values

Since there is an even number of values, the median is the mean of the two middle values.

i.e. median $= \dfrac{50 + 51}{2} = 50.5$

Spread

This section deals with the following four measures of spread:
* range
* inter-quartile range
* variance
* standard deviation.

A very elementary data set (Table 3.1) has been used throughout, simply to lay out the bare bones of each calculation.

Table 3.1 *The heights of nine women*

Name	Height (m)
Roberta	1.56
Alice	1.62
Rajvinder	1.71
Jenny	1.58
Fiona	1.67
Sumita	1.66
Hilary	1.60
Marcie	1.59
Sue	1.61

Five-figure summary

Before looking at the range and inter-quartile range, it is necessary to scan these nine numbers in order to pick out five of the key figures. This is easiest to do when the numbers are sorted in order of size, as shown in Table 3.2.

* The lower extreme value (E_L) is the height of the shortest person, Roberta, with a height of 1.56 m.

* The upper extreme value (E_U) is the height of the tallest person, Rajvinder, with a height of 1.71 m.

* The median value (Md) lies half way through the values in Table 3.2. This corresponds to Sue's height, so Md = 1.61 m.

* The lower quartile (Q_L) lies a quarter of the way through the values in Table 3.2. To put this another way, the lower quartile lies half way through the lower half (i.e. covering the five shortest people) of the batch of data. Marcie is third of these five women, so Q_L = 1.59 m.

* The upper quartile (Q_U) lies three-quarters of the way through the values in Table 3.2. By the same reasoning as before, this will correspond to the median height of the five tallest women. This will be Sumita's height, so Q_U = 1.66 m.

These five numbers provide a useful summary of a batch of data and are highlighted in Table 3.2.

Table 3.2 *The heights of nine women, sorted in order of size*

Name	Height (m)	Five key figures
Roberta	1.56 ⟶	The lower extreme value (E_L)
Jenny	1.58	
Marcie	1.59 ⟶	The lower quartile (Q_L)
Hilary	1.60	
Sue	1.61 ⟶	The median height (Md)
Alice	1.62	
Sumita	1.66 ⟶	The upper quartile (Q_U)
Fiona	1.67	
Rajvinder	1.71 ⟶	The upper extreme value (E_U)

Boxplot

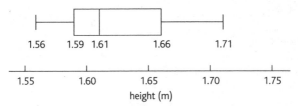

Figure 3.1 *Boxplot drawn from the highlighted values in Table 3.2.*

In general, then, the significant parts of the boxplot are as follows.

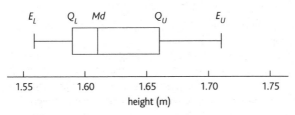

Figure 3.2 *Boxplot showing where the five key values are placed.*

The central rectangle which marks out the two quartiles is called the 'box', while the two horizontal lines on either side are the 'whiskers'. Just by observing the size and balance of the box and the whiskered components we can gain a very quick and useful overall impression of how the batch of data is distributed. Also, drawing two boxplots one above the other can provide a powerful means of comparing two distributions.

Range

The range is the simplest possible measure of spread and is the difference between the upper extreme value (E_U) and the lower extreme value (E_L). Returning to the height example, the tallest person is Rajvinder at 1.71 m and the shortest is Roberta at 1.56 m. Thus:

$$\text{Range} = E_U - E_L$$

so range = 1.71 m – 1.56 m = 15 cm.

One disadvantage of the range as a measure of spread is that it is strongly affected by an extreme or untypical value. For example, it only takes one extremely tall or extremely short person in the sample to have an enormous effect on the value of the range. This problem can be overcome by choosing to measure the range, not between the two extreme values, but between two other values on either side of the middle value. And what could be two more appropriate values to choose than the upper and lower quartiles!

Inter-quartile range (DQ)

The inter-quartile range, sometimes known as the inter-quartile deviation, is, as you might expect, the difference between the two quartiles. In this example it is the difference between the heights of Sumita and Marcie.

$$\text{The inter-quartile range, } dq = Q_U - Q_L$$

so inter-quartile range = 1.66 m – 1.59 m = 7 cm.

The inter-quartile range therefore contains the middle half of the batch values, which is, of course, the central box part of the boxplot.

A difficulty with any measure of spread which is tied to the units of the batch values is that it makes comparisons with the spread of other batches of data very misleading. For example, how might you compare the spread of a group of people's heights, measured in metres, with the spread of their weights, measured in kg? And would the spread of heights suddenly increase by a factor of 100 if the data values were converted from metres to centimetres?

Variance (σ^2) and Standard deviation (σ)

The method of calculating the variance is described below.

(i) Calculate the overall mean.

(ii) Subtract the mean from each value in the batch. This produces a set of deviations, d (where $d = X - \bar{X}$ for each value of X). Take the square of these deviations, d^2.

(iii) Find the mean of these squared deviations $\frac{\Sigma d^2}{n}$. The result is the 'variance'.

(iv) If you wish to find the standard deviation, take the square root of the variance.

Now, here are the formulas for these two measures of spread.

$$\text{Variance, } \sigma^2 = \frac{\Sigma d^2}{n}$$

$$\text{Standard deviation, } \sigma = \sqrt{\frac{\Sigma d^2}{n}}$$

Finally, Table 3.3 shows how to calculate the variance and standard deviation, using the height data.

Table 3.3 *Calculation of the variance and standard deviation*

Name	Height (X) metres	Deviations (d)	Squared deviations (d^2)
Roberta	1.56	−0.062	0.003 844
Jenny	1.58	−0.042	0.001 764
Marcie	1.59	−0.032	0.001 024
Hilary	1.60	−0.022	0.000 484
Sue	1.61	−0.012	0.000 144
Alice	1.62	−0.002	0.000 004
Sumita	1.66	0.038	0.001 444
Fiona	1.67	0.048	0.002 304
Rajvinder	1.71	0.088	0.007 744
Sum	14.60	0	0.018 756
Mean	1.622	0	0.002 084
			0.045 651

stage (ii) — (points to squared deviations 0.001 764 … 0.000 144)

stage (iii), variance — (points to 0.007 744)

stage (i) — (points to Mean 1.622)

stage (iv), standard deviation — (points to 0.045 651)

So, variance = 0.002 084 m² and standard deviation = 0.045 651 m. If this were a practical example, these results would, of course, be rounded to an appropriate number of decimal places, depending on the circumstances of the question.

The procedure for calculating the variance and standard deviation is slightly more complicated when applied to data that have been organized into a frequency or percentage table. However, the method is exactly equivalent to the approach used when calculating a weighted arithmetic mean; namely, you have to remember to multiply each value (in this case, each squared deviation) by its corresponding weight, f, before finding the sum of the squared deviations. The formulas for finding a weighted variance and standard deviation are given below.

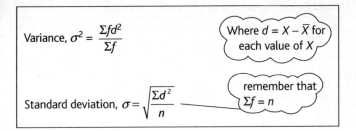

Variance, $\sigma^2 = \dfrac{\Sigma f d^2}{\Sigma f}$

Where $d = X - \bar{X}$ for each value of X

Standard deviation, $\sigma = \sqrt{\dfrac{\Sigma d^2}{n}}$

remember that $\Sigma f = n$

There are other formulas for calculating the variance and standard deviation which will produce the same results but which make for a slightly easier calculation. For reasons of space these are not explained here, but increasingly these short-cut methods are becoming irrelevant as more and more users are performing such calculations on machines.

Finally, it is worth noting that of all the measures of spread that are available, the standard deviation is the one most widely used.

4

choosing a sample

A remark that you sometimes hear when the results of some survey are being discussed is, 'Well, nobody asked me!'. The whole point of surveying opinions by taking a sample is that, provided the sample is carefully selected so as to be representative of the entire population, it should provide an accurate overview without having to go through the cost and effort of asking everyone's opinion. An analogy that students find useful here is that of a cook making a large pot of soup. In order to check its flavour, she will first give the pot a good stir (to ensure that that it is thoroughly and evenly mixed) and then sample a (representative) spoonful. Doing this avoids the need for her to consume the entire potful in order to make a sensible judgement about its flavour!

In general, we choose a sample in order to measure some property of the wider population from which it was taken. This may be in order to discover certain information about a single population (the extent of childhood illness, how many cigarettes people smoke, and so on), or the sampling may be part of an investigation into whether two or more populations are measurably different (are there regional differences in childhood illness, do boys smoke more than girls, and so on). A 'good' sample is one which fairly represents the population from which it was taken.

The sampling frame

When talking about sampling in statistics, it is useful to separate out three related but distinct levels of terminology. Starting at the top level first, these are known as the 'population', the 'sampling frame' and 'sample'. Figure 4.1 below suggests that the sampling frame is (often) a subset of the population and the sample, in turn, is a subset of the sampling frame.

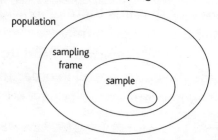

Figure 4.1 *Population, sampling frame and sample.*

* **Population** – in statistics, this term 'population' isn't restricted in meaning to refer just to a population of animals or humans but is used to describe any large group of things that we are trying to measure. This might be a precise measure of the length of each item of a particular manufactured component, or the weight of each bag of crisps coming off a production line, or the life in hours of

a particular brand of light bulbs. In general, manufacturers need to monitor their production process to ensure that their grommets or their bags of crisps, etc. are continuing to meet the required standard. Also, governments expect to collect detailed information about many aspects of the human populations for whom they provide services and from whom they collect money. It would be too expensive and time-consuming to measure each grommet or weigh each bag of crisps or interview each person in the population, so they all usually resort to some sort of sampling technique. The size of the population, i.e. the number of items in the population, is usually denoted by an upper case N.

* **Sampling frame** – this is a list of the items from which the sample is to be chosen. Often it is the case that the sample is taken directly from the population, in which case the sampling frame would simply be a complete list cataloguing each item in the population. Sometimes, however, the population is simply too large and complicated for each element in it to be itemized accurately. For example, if the government wishes to estimate the national acreage of land growing oilseed rape, a suitable sampling frame would be a list of farms in the country. However, it is a major undertaking to list all the farms in the UK and it is likely that a number of small-holdings will be omitted.

* **Sample** – this, in turn, is the representative subset of the sampling frame which is chosen as fairly as possible to represent the entire population. The term 'sample size', usually denoted by a lower case letter 'n', refers to the number of items in the sample. Where more than one sample is taken, it is worth stressing that 'n' is the number of items selected in a particular sample and not the number of samples taken. The (rare) situations where the entire population is 'sampled', i.e., where $n = N$, is known as a 'census'.

Random sampling

There are several techniques for choosing a representative sample, of which probably the best known is *random sampling*. Its basic definition is as follows.

> A random sample is one in which every item in the population is equally likely to be chosen.

With random sampling, there also needs to be some sort of random-generating process used to select each item. Suppose, for example, we wished to select a sample of size 50 (sample size, $n = 50$) from a population consisting of a sampling frame of size 1000 (i.e. $N = 1000$). The usual approach would be to allocate a number to each item in the sampling frame (from 1 to 1000 or perhaps from 000 to 999) and then find some way of randomly generating 50 numbers from within this range. The population items which had been allocated these numbers then become the 50 items which form our sample.

There are two common methods of random sampling – sampling with replacement and without replacement. Replacement means simply putting an item back into the sampling frame after it has been selected. Although replacement is sometimes thought of as being a more correct form of random sampling, in most practical sampling situations items are not replaced after they have been selected.

Generating random numbers

Coins and dice are quite good random number generators in that their symmetrical shapes ensure fairly random sequences. However, for selecting samples of say, 15 or 20 or more from larger populations, the whole thing becomes a bit of a chore. You may be thinking that it would all be much easier if the throws of the die had already been done for you and the scores were set out in a table. In fact, this is exactly what a *random number table* provides, although the numbers are randomized in the range 0 to 9, rather than 1 to 6.

An alternative to using a random number table is to generate random numbers from a computer or calculator. Many scientific calculators possess a key marked something like 'Rand' or 'Ran#' which, typically, will produce a random decimal number in the range 0 to 1. Alternatively it may be possible to use this facility to generate whole numbers randomly within a specified range.

Human error and sampling error

I reckon the light in my 'fridge' must stay on all the time. I've checked hundreds of times and every time I open the door to have a look it is still on!

One of the dilemmas for any researcher is that he or she never really knows the extent to which their research observations affect or disturb the objects or people being observed. As with the 'opening the "fridge" door' example above, all surveys involve opening some sort of door on people's lives. Observers participate in this process and can never fully know the extent to which their presence has affected the phenomena that are being presented to them. All they can do is attempt to monitor their findings against those collected in a different way and then try to account for any differences. Most of all, the observer needs to be sensitive to these questions and make every attempt to tread lightly.

There are two main types of error that surround sampling. The first is the sort introduced by a badly designed sampling procedure, and some examples of these have been given earlier in the chapter. This might be called 'human error'. Sometimes error is deliberately and unfairly introduced into the sampling process in order to distort the findings. Examples of this are the original polls which were used to compile the pop music charts. The first singles chart, known as the Record Hit Parade, was released in 1952 and was published by the *New Musical Express*. (Before that, the charts were compiled from sheet music sales.)

In those early days, a small number of record shops were selected and sampled for their record sales over the previous week. The problem was that the sample size was small and, once the identity of a 'chart shop' became known, agents would descend with wads of fivers to buy particular records and thereby bump up their chart ratings. This problem was tackled by increasing the number of 'chart shops', and today the Gallup polling agency is responsible for compiling the charts samples from around 1000, or roughly one in four, of all music outlets.

Phone-in polls are a popular means by which some radio stations compile record charts, but these are even more open to distortion than sampling sales from record outlets. In 1986, when Capital Radio decided to compile the all-time Top 100 records, based on a listeners' phone-in poll, five Bros records featured in the chart. This probably says something about the promotional zeal of the listening 'Brosettes' and the fact that a phone call or ten costs less than the price of a single, particularly if mummy and daddy are footing the telephone bill.

Let us now turn to the second form of error. Due to natural variation, all sampling is inherently prone to error and this variation is an inevitable part of the random sampling process. It is this type of error which is referred to as 'sampling error'. Although sampling error cannot be eliminated, it is possible to quantify how much variation to expect under different conditions, and this allows us to gain some measure of the confidence with which we can make predictions about the population.

The larger the sample you take, the greater your confidence will be that the sample result is accurate, so therefore the narrower will be your confidence interval. This may appear to be slightly counterintuitive (the notion of high confidence being linked to 'narrow' confidence) but it may be helpful to think of the confidence interval as a 'tolerance' – a low error tolerance is more comfortably associated with a high degree of confidence in the population estimate. The best way to be convinced about this relationship between confidence and sample size is to do a simple sampling experiment based on samples of different size. If you then plot the sample means you will find that the means which were calculated from the smaller samples are spread out more widely than the means taken from large samples.

5

collecting information

When conducting a statistical investigation, it is helpful to plan in advance the steps that you need to go through. The first step is to clarify your central question; the more purposeful and clearly formulated your question, the easier it will be to make decisions later. Step 2 is data collection. A guiding principle here is that data generation costs time and money, so try to avoid re-inventing the wheel. Before you rush out and try to generate your own data, therefore, make sure that someone else hasn't already done it for you. Data that have already been collected by someone else and are awaiting your attention are known as *secondary source data* or sometimes simply as secondary data. Alternatively, you may have to carry out an experiment or survey yourself, in which case this would involve you in generating *primary source data* (also known as primary data).

Secondary source data

Historically, the term 'statistics' derives from 'state arithmetic'. For thousands of years, rulers and governments have felt the need to measure and monitor the doings of their citizens. From raising revenue in the form of taxes to counting heads and weapons in times of war, the state has always been the major driving force in collecting and publishing statistical information of all kinds. And, as you will see from this section, they are still at it! Just a few of the most useful secondary sources of data are described here, most of which are from government sources.

Let us assume that you have identified a particular question for investigation. As has already been suggested, the first task is to check whether any suitable secondary source data are available. You enter your local library, but where do you look? A good starting place is usually the latest edition of *Social Trends*. This annual publication of the Government Statistical Service is prepared by the Office for National Statistics (ONS) and published by HMSO. As well as *Social Trends*, the best known statistical publications from the ONS are *Economic Trends* and *The Monthly Digest*. They also produce monthly Business Monitors and ONS Bulletins and have a wide range of computerized data on disk. A lot of this information is now also available online at www.statistics.gov.uk.

Social Trends

This excellent publication is crammed with interesting tables and charts about a wide variety of social and economic issues. At the time of writing, the data it provides are presented in the following 13 chapters.

1 Population
2 Households and Families
3 Education
4 Employment
5 Income and Wealth
6 Expenditure and Resources
7 Health and Personal Social Services

  8 Housing

  9 Environment

  10 Leisure

  11 Participation

  12 Law Enforcement

  13 Transport.

Depending on the particular original source being quoted, the data are based on UK, Great Britain or England and Wales but comparisons are given over time (anything from, say, 10 to 30 years) and against selected other countries (particularly EU countries). Each chapter ends with a bibliography listing the main sources used. These references are themselves useful should you not find the relevant item of secondary source data in the *Social Trends* publication directly. The appendix to *Social Trends* lists the major surveys used, and some of the best known of these are listed below.

The *Family Expenditure Survey* (FES), which investigates, annually, the income and spending patterns of a sample of around 7000 households in the UK.

The *Survey of Retail Prices*. This is carried out monthly by the Office for National Statistics and provides about 150 000 prices used in calculating the Retail Prices Index (RPI). (*The Employment Gazette*, HMSO, provides a useful, detailed and up-to-date summary of these data each month.)

The *New Earnings Survey* (NES), which collects information on an annual basis on the earnings of about 180 000 people. (This information is published annually in the *New Earnings Review*.)

The *British Social Attitudes Survey*. This annual survey solicits the opinions and attitudes of between three and four thousand individual adults each year, and the results are written up in the annual report, *British Social Attitudes*.

The *General Household Survey* collects a wide range of information on households from over 10 000 addresses, resulting in an annual publication of the same name.

The *Census of Population* takes place every 10 years (in 1991, 2001, 2011, etc.) and is based on a restricted amount of information about personal details and the nature of households of all citizens of the UK. It is used as a source of reliable statistics for small groups and areas throughout the country so that government and others can plan housing, health, transport, education and other services. The 2001 census data were collected from some 24 million households. Local area statistics from the census were published separately for each county and all the key results of each county are also available in machine readable format.

One frustration with secondary data is that they are rarely in exactly the form that you want. You may think that you have collected information on gross earnings in the UK only to read the small print and discover that the figures refer to Great Britain only or that part-time workers have been excluded and you need the inclusive figure for purposes of comparison. Fortunately there is a simple remedy for such problems – it is called 'hard graft'! You need to be prepared to put in the time and effort checking out all the small print of any secondary data that you propose to use and making sure that they really are the figures that you want.

As often as not you will need to chase up some additional data from elsewhere.

Primary source data

It may be that your research question is concerned with an issue that is too specific or too local for there to be suitable secondary source data available, in which case you may decide to collect your own data. This may require accurate measurement that involves carrying out some sort of scientific experiment, or perhaps designing a questionnaire and conducting a sample survey.

Questionnaire design

Exercise 5.1 A questionable questionnaire

The following questionnaire, surveying people's health practices and attitudes, is based on examples given in the book Surveys in Social Research, by D.A. de Vaus (George Allen & Unwin, 1986). Each question is something of a lemon in the way it is worded.

Have a go at answering the questionnaire and then give some thought to the sorts of biases and inaccuracies it might produce.

A survey on health
 1 How healthy are you?
 2 Are the health practices in your household run on matriarchal or patriarchal lines?
 3 Has it happened to you that over a long period of time, when you neither practised abstinence nor used birth control, you did not conceive?
 4 How often do your parents visit the doctor?
 5 Do you oppose or favour cutting health spending, even if cuts threaten the health of children and pensioners?
 6 Do you agree or disagree with the following statement? 'Abortions after 28 weeks should not be decriminalized'.
 7 Do you agree or disagree with the government's policy on the funding of medical training?
 8 Have you ever murdered your grandmother?

[Comments below]

As you will have discovered, this is a pretty hopeless questionnaire but there are some important general principles that can be learned from it.
 1 There is no way of knowing how to answer a question like this. There needs to be included either a clear set of words to choose from or a scale of numbers on which to rate your perception of how healthy you feel.
 2 Don't ask questions that people won't understand.
 3 Simplify the wording and sentence structure so that the question is clear and unambiguous without being trite or patronizing. Don't use undefined time spans (like 'a long period of time') and avoid using double negatives.
 4 Avoid double-barrelled questions like this where two (or more) individuals have been collapsed into one category (in this case, 'parents'). Also, where a quantitative answer is

required, give the respondent a set of categories to choose from (e.g. 'every day', 'roughly once a week', etc.). Finally, avoid gathering information from someone on behalf of someone else. They probably won't know the exact details and anyway it is none of their business.

5 This is a leading question that is pushing for a particular answer and will produce a strong degree of bias in the responses.

6 Again, avoid double negatives as they are hard to understand. This could be reworded as 'Abortions after 28 weeks should be legalized'.

7 Most people will not know what the government's policy is on the funding of medical training, so don't ask questions that people don't know about without providing further information.

8 This could be classed as a direct question on a rather sensitive issue. If you really must ask this question, there are ways of asking it more tactfully. According to de Vaus, there are four basic gambits.

(a) The 'casual' approach – 'Do you happen to have murdered your grandmother?'

(b) The 'numbered card' approach – 'Will you please read off the number of this card which corresponds with what became of your grandmother?'

(c) The 'everybody' approach – 'As you know, many people have been killing their grandmothers these days. Do you happen to have killed yours?'

(d) The 'other people' approach – 'Do you know any people who have murdered their grandmothers?' Pause for a reply and then ask, 'Do you happen to be one of them?'

There are several other aspects to designing a good questionnaire that have not emerged from this one. First and foremost, only ask questions that you want and need to know the answer to and that will provide you with information which you intend to use. Most people have only a limited capacity for filling in forms and answering questions, so don't use up their time and goodwill asking unnecessary questions. Secondly, you will need to decide whether to collect quantitative or qualitative data or a combination of the two. In making this choice, it is helpful to bear in mind how you intend, subsequently, to process the information that you collect. Qualitative information will be difficult to codify and summarize but it may allow options and opinions to be expressed that you would never otherwise hear. Quantitative data are easier to process but have the disadvantage that the range of responses is restricted to the ones you thought of when designing the questionnaire. A helpful solution to this dilemma is to offer a set of pre-defined categories but allow the respondent to add a final option of their own in the space provided, thus:

'Other, please specify' _____

Also, with questions where opinions are being sought, there may be issues on which people will have no opinion, so the option of recording 'don't know' or 'no opinion' should be offered in the range of choices.

6

spreadsheets to the rescue

Increasingly these days, data handling is done on a computer and the most common application for this purpose is a spreadsheet. If you have access to a computer with a word processor, the chances are that it already has a spreadsheet application available to you as part of a wider 'office software' package. The most commonly used spreadsheet application is Excel, which is part of Microsoft Office (popular but expensive) but there are many good alternatives, such as Google Spreadsheets.

For simplicity, the examples provided in this chapter are based on very small data sets. However, one of the great strengths of spreadsheets is that, whether your columns of data contained 10 or 100 or 1000 items of data, the commands would have been equally simple to enter and the time taken for your spreadsheet to make the calculation would still be only fractions of a second.

What is a spreadsheet?

A spreadsheet is a computer tool that provides a way of laying out data in rows and columns on the screen. The rows are numbered 1, 2, 3, etc. while the columns are labelled with letters, A, B, C, etc. Typically a spreadsheet might look something like the grid in Figure 6.1, except that rather more rows and columns are visible on the screen at any one time.

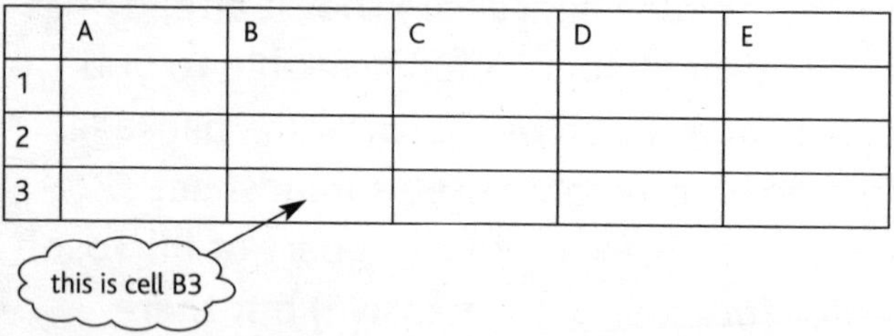

	A	B	C	D	E
1					
2					
3					

this is cell B3

Figure 6.1 *Part of a spreadsheet.*

Each 'cell' is a location where information can be stored. Cells are identified by their column and row position. For example, cell B3 is indicated in Figure 6.1, and is simply the cell in the third row of column B.

The sort of information that you might wish to put into each cell will normally fall into one of three basic types – numbers, text or formulas.

* *Numbers* These can be either whole numbers or decimals (e.g. 7, 120, 6.32, etc.)
* *Text* These are symbols or words (e.g. the headings of a row or a column of numbers will be entered as text)
* *Formulas* The real power of a spreadsheet lies in its ability to handle formulas. A cell which contains a formula will display the result of a calculation based on the current values contained in other cells in the spreadsheet (e.g. an average or a total or perhaps something more complicated such as a bill made up from different components). If these other component values are subsequently altered, the

formula can be set up so that it automatically recalculates on the basis of the updated values and immediately gives the new result.

Why bother using a spreadsheet?

A spreadsheet is useful for storing and processing data when repeated calculations of a similar nature are required. Next to wordprocessing, a spreadsheet is the most frequently used computer application in business, particularly in the areas of budgeting, stock control and accounting. Spreadsheets are also being used increasingly by householders to help them to solve questions that crop up in their various roles – as shoppers, tax payers, bank account holders, members of community organizations, hobbyists, etc. They can be used to investigate questions such as:

* how much will this journey cost for different groups of people?
* is my bank statement correct?
* which of these buys is the best value for money?
* what is the calorific content of these various meals?
* what would these values look like sorted in order from smallest to biggest?
* how can I quickly express all these figures as percentages?

The reason a spreadsheet is such a powerful tool for carrying out repeated calculations is that, once it has been set up properly, you simply perform the first calculation and then a further command will complete all the other calculations automatically. Another advantage of a spreadsheet over pencil and paper is its size. The grid that appears on the screen (part of which was illustrated in Figure 6.1) is actually only a window on a much larger grid. In fact, most spreadsheets have available hundreds of rows and columns, should you need to use them. Movement around the spreadsheet is also quite straightforward – you can use certain keys to move between adjacent cells or to a particular cell location of your choice.

Using a spreadsheet

In this section you will be guided through some simple spreadsheet activities so, if possible, switch on the computer and let's get started.

There are many spreadsheet packages on the market and fortunately their mode of operation has become increasingly similar in recent years. However, your particular package may not work exactly as described here so you may need to be a little creative as you try the activities below.

A very useful feature of any spreadsheet is that it can add columns (or rows) of figures. This is done by entering a formula into an appropriate cell. For most spreadsheets, formulas are created by an entry starting with '='.

Exercise 6.1 Shopping list

Load the spreadsheet package and, if there isn't already a blank sheet open, create one by clicking on File and selecting New.

Using the mouse, click on cell A1 and type in 'Milk'. Notice that the word appears on the 'formula bar' near the top of the screen. Press Enter (the key may be marked Return or have a bent arrow pointing left) and the word 'Milk' is entered into cell A1.

Using this method, enter the data below into your spreadsheet.

	A	B
1	Milk	0.74
2	Bread	0.88
3	Eggs × 12	2.65

Exercise 6.2 Finding totals using a formula

The formula for adding cell values together is = sum( ). Inside the brackets are entered the cells or cell range whose values are to be added together.

Click in cell B4 and enter: = sum(B1:B3)

Press Enter and the formula in cell B4 produces the sum of the values in cells B1 to B3. Now enter the word 'TOTAL' into A4.

Your spreadsheet should now look like this.

	A	B
1	Milk	0.74
2	Bread	0.88
3	Eggs × 12	2.65
4	TOTAL	4.27

Let's suppose that you gave the shopkeeper £10 to pay for these items. How much change would you expect to get? Again, this is something that the spreadsheet can do easily. Calculations involving adding, subtracting, multiplying and dividing require the use of the appropriate operation keys, respectively marked +, −, * and /. (Note that the letter 'x' cannot be used for multiplication: you must use the asterisk, *). You will find these keys, along with the number keys and '=', conveniently located on the numeric keypad on the right-hand side of your keyboard. (These keys are all also available elsewhere on the keyboard, but you may have to hunt them down, which takes time.)

Exercise 6.3 Calculating change from £10

Enter the word 'TENDERED' in cell A5, the number '10' in cell B5 and the word 'CHANGE' in A6. Now enter a formula into B6 which calculates the change.

Remember that the formula must begin with an equals sign and it must calculate the difference between the value in B5 and the value in B4.

[Comments below]

The required formula for cell B6 = B5 − B4.

Let's now try a slightly more complicated shopping list, this time with an extra column showing different quantities. Enter the data below into your spreadsheet, starting with the first entry in cell A8.

	A	B	C	D
8	DESCRIPTION	QUANTITY	LIST PRICE (£)	COST (£)
9	Pens black	24	0.87	
10	Folders	15	0.65	
11	Plastic tape	4	1.28	
			TOTAL	

In order to calculate the overall total cost, you must first work out the total cost of each item. For example, the total cost of the black pens is 24 × £0.87.

Enter into cell D9 the formula: = B9*C9

This gives a cost, for the pens, of £20.88. You could repeat the same procedure separately for the folders and the plastic tape but there is an easier way, using a powerful spreadsheet feature called 'fill down'. You 'fill down' the formula currently in D9 so that it is copied into D10 and D11. The spreadsheet will automatically update the references for these new cells.

Click on cell D9 (which currently displays 20.88) and release the mouse button. Now move the cursor close to the bottom right-hand corner of cell D9 and you will see the cursor change shape (it may change to a small black cross, for example). With the cursor displaying this new shape, click and drag the mouse to highlight cells D9–D11 and then release the mouse button. The correct costs for the folders (£9.75) and the plastic tape (£5.12) should now be displayed in cells D10 and D11 respectively. Now click on cell D10 and check on the formula bar at the top of the screen that the cell references are correct. Repeat the same procedure for cell D11. You should find that they have indeed automatically updated – magic!

What else will a spreadsheet do?

This chapter can only really scratch the surface of what can be done with a spreadsheet. As has already been mentioned, the tiny data sets used here are merely illustrative and don't properly reveal the power of a spreadsheet.

Once data have been entered into a spreadsheet, there are many options available for seeking out some of the underlying patterns. For example, columns or rows can be re-ordered or sorted either alphabetically or according to size. As you have already seen, column and row totals can be inserted. Also, you can easily divide one column of figures by numbers in an adjacent column to calculate relative amounts or to convert a set of numbers to percentages.

Spreadsheet commands can include a number of different mathematical functions. First, there are the four operations +, −, × and ÷ which can be found directly on the computer keyboard. In terms of statistical work, there are many other functions contained within one of the spreadsheet's menu options (or they can be typed in directly, letter by letter). These will enable you to select some or all of the data and find various summaries such as the mean, median, maximum value, standard deviation, and much more, at the touch of a button.

7

regression: describing relationships between things

In statistics, exploring the relationship between two things usually involves collecting paired data and plotting them in the form of a scatter plot. Regression is the procedure of fitting a trend line or curve to a scatter of points in such a way that it best describes the general nature of the relationship.

Depending on the pattern of the points, a choice must be made as the best model to choose: perhaps linear, quadratic or something else. In this chapter just linear regression is considered. The technical side of regression is to work out where the best-fit line is located by finding its equation.

The regression line is useful because it allows predictions to be made. *Interpolation* refers to predictions made within the known range of data. The riskier procedure of *extrapolation* involves extending the regression line beyond this range in order to make predictions.

Paired data

Data collection generally takes place from a sample of people or items. Depending on the purpose of the exercise, we often take just *one single* measure from each person or item in the sample. For example, a nurse might wish to take a group of patients' temperatures, or a health inspector might measure the level of pollution at various locations along a river. These sorts of sample will consist of a set of single measurements which should help to provide a picture of the variable in question. So, the nurse could use the temperature data to gain some insight into the average temperatures of patients in hospital and how widely they are spread. Similarly, the health inspector could use the pollution data to establish the typical levels of pollution present in the river, and also see to what extent the levels of pollution vary from place to place.

Sometimes, however, data can be collected *in pairs*. Paired data allows us to explore something quite different than is indicated by the examples above. Using paired data, we can look at the possible *relationship* between the two things being measured. To take a simple example, look at the two graphs, Figure 7.1, which show the typical amounts of sunshine and rainfall over the year in the Lake District.

(a) Generally, the sunniest months are in the middle of the year, around May and June, while the months at the beginning and end of the year have the least sunshine.

(b) Rainfall is least in the middle of the year, around May and June, while the months at the beginning and end of the year have most rainfall.

(c) It seems clear from (a) and (b) above that there is a relationship between rainfall and sunshine. In general, the months which have the most sunshine tend to have the least rainfall, and vice versa. The pattern in this relationship can be spelt out more clearly when drawn on a scattergraph (Figure 7.2).

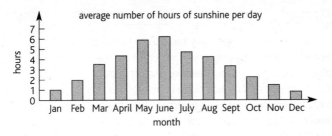

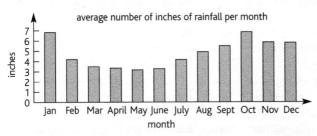

Figure 7.1 *Weather in the Lake District.*
Sources: various.

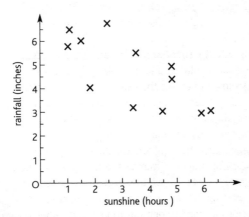

Figure 7.2 *Scattergraph showing the data from Figure 7.1.*

If you are in any doubt about how to plot the points, you can check your scattergraph with the one given in Chapter 1. As you should see from your scattergraph, the pattern of points is fairly clear. The points seem to lie roughly around a straight line which slopes downwards to the right. This bears out the conclusions that high sunshine tends to be associated with low rainfall. The main ideas of this chapter and the one which follows it are all about *interpreting more precisely the sort of patterns that can be found on a scattergraph*. A key point to remember, however, is that these ideas involve investigating relationships between two things and therefore throughout this chapter we will be dealing with *paired data*.

The 'best-fit' line

The 'best-fit' line or *regression line*, as it is sometimes called, is the best line which can be made to fit the points which it passes through. However, a straight line may be rather too crude a description of the trend. In the world of economic and industrial forecasting, for example, curves are commonly used and these provide a necessary degree of subtlety and accuracy. For instance, economic data are often subject to seasonal fluctuations (people tend to spend more money on household goods around December and January and less in the summer). Fitting straight lines to these sorts of data would be to ignore this seasonal factor and result in highly inaccurate predictions. But for the purposes of this chapter, we will look only at methods using straight lines. This is called *linear* (i.e. straight line) regression.

In order to define a straight line exactly, two pieces of information are required. By convention, these are usually the *slope* of the line, and the point on the graph where it crosses the vertical axis (known as the *intercept*). The slope can be either a positive or a negative number. Lines which slope from bottom left to top right will have a positive slope – i.e. the value of Y will increase as the value of X increases. Lines which have a negative slope run from top left to bottom right.

In this example, Rainfall is on the vertical axis and Sunshine on the horizontal axis (Figure 7.3). The conventional labels for these axes are, respectively, Y and X. In general, if we are seeking to find the linear regression line, the equation of the straight line required will be of the form:

$$Y = a + bX$$

where Y represents Rainfall (measured in inches per month)
       X represents Sunshine (measured in hours per day)

and a and b are the two pieces of information required to define the line,

a = intercept
b = the slope.

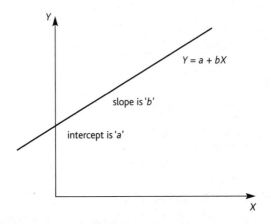

Y

$Y = a + bX$

slope is 'b'

intercept is 'a'

X

Figure 7.3 *The regression coefficients* a *and* b.

There are several other methods for finding the values of a and b, the simplest of which is to feed the raw data directly into a suitable calculator (or computer) and obtain them at the press of a button. When all the data have been entered, you then simply press the keys marked 'a' and 'b' to discover the values for the intercept and slope of the best-fit line.

Making predictions

Essentially, this best-fit line is a mathematical summary of what you think might be the relationship between the two variables in question. However, what is more interesting than calculating it is to use and interpret it. What we can now do is to use the line to make predictions about other pairs of values, perhaps by extending the line or by making guesses about points on the line where data are absent.

However, it is worth stressing that these really are only guesses which carry all sorts of assumptions that may not hold in practice.

Predictions which lie within the range of the sample, are called *interpolation* (literally 'within the points'). Of course, even interpolated estimates may be extremely inaccurate for a number of reasons – the regression line is only an average, after all, and merely gives rough predictions.

The dangers of inaccuracy increase when making predictions which involve having to extend the line beyond the range of values in the sample. Such predictions are known as *extrapolation* (literally going 'beyond the points').

Calculating the regression coefficients using the formula

An accurate method for finding the regression coefficients is to perform a rather complicated calculation, using the formulas given opposite.

To find the slope, b, you first need to calculate the following intermediate values:

n	$\rightarrow$	this is the number of pairs of values (in this example, $n = 12$)
ΣXY	$\rightarrow$	the sum of the 12 XY products
ΣX	$\rightarrow$	the sum of the X values
ΣY	$\rightarrow$	the sum of the Y values
ΣX^2	$\rightarrow$	the sum of the squares of the X values

These intermediate values are then substituted into the following exciting-looking formula (which will not be justified but simply stated here).

$$b = \frac{n\Sigma XY - \Sigma X\Sigma Y}{n\Sigma X^2 - (\Sigma X)^2}$$

Having calculated the value for b, finding a follows more easily. The mid-point of all the points on the scattergraph is denoted by the coordinates $(\overline{X}, \overline{Y})$. (*Note:* The sigma ($\Sigma$) and bar ($\overline{X}, \overline{Y}$) notations used above are explained in Chapter 1.)

An important property of the best-fit line is that it passes through this mid-point. That being so, we can substitute these coordinates, $(\overline{X}, \overline{Y})$, into the equation of the line $Y = a + bX$. This gives:

$$\overline{Y} = a + b\overline{X}$$

Rearranging this equation produces the following:

$$a = \overline{Y} - b\overline{X}$$

from which we can find the value of a.

8

correlation: measuring the strength of a relationship

Correlation measures the strength of association between two things. A useful visual indication of correlation is the degree of the scatter around the best-fit line when the data are plotted on a scatter plot. Positive correlation shows a clear pattern of points running from bottom left to top right, while negative correlation shows a pattern of points running from top left to bottom right.

There are two common measures of correlation: the Pearson coefficient (r) and, for ranked data, the Spearman coefficient (r'). Both measures produce values lying in the range −1 to 1. A coefficient value of −1 means perfect negative correlation, while a coefficient value of 1 means perfect positive correlation. When r or r' take a value of around zero, there is no correlation between the two factors in question. Finally, note that strong correlation is no proof of whether or not the relationship in question is one of cause and effect.

This chapter builds on the ideas of regression which were the subject of the previous chapter and explores the degree of confidence we can have that the relationship under consideration really does exist. In other words, we are interested in knowing about the strength of the relationship.

Scatter

As has already been suggested in the previous chapter, the regression line is useful as a statement of the underlying trend but tells us nothing about how widely scattered the points are around it. The degree of scatter of the points, or the *correlation*, is a measure of the strength of association between two variables. For example, perfect positive correlation will look like Figure 8.1.

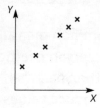

Figure 8.1 *Scattergraph showing perfect positive correlation.*

Perfect negative correlation will result from data which are plotted as in Figure 8.2.

In practice, most scattergraphs portray relationships with correlations somewhere between these two extremes. Some examples are shown in Figures 8.3 to 8.5.

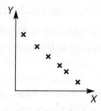

Figure 8.2 *Scattergraph showing perfect negative correlation.*

Figure 8.3 *Strong positive correlation.*

Figure 8.4 *Weak negative correlation.*

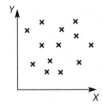

Figure 8.5 *Zero correlation.*

Product-moment correlation coefficient

Although it is useful to be able to gauge the strength of the relationship simply by looking at the scattergraph, a more formal method based on calculation is available which gives a numerical value for the degree of correlation. The product-moment coefficient of correlation (sometimes known as Pearson's coefficient) is calculated on the basis of how far each point lies from the 'best-fit' regression line. It is denoted by the symbol r.

The formula for r, the product-moment correlation coefficient is:

$$r = \frac{S_{XY}}{S_X S_Y}$$

where X is the variable on the horizontal axis
and Y is the variable on the vertical axis

S_{XY} is the covariance, given by the formula $= \dfrac{\sum XY}{n} - \overline{X}\,\overline{Y}$
S_X is the standard deviation of $X = \sqrt{\dfrac{\sum X^2}{n} - \overline{X}^2}$
S_Y is the standard deviation of $Y = \sqrt{\dfrac{\sum Y^2}{n} - \overline{Y}^2}$

The formula has been devised to ensure that the value of r will lie in the range – 1 to 1. For example:

$r = -1$ means perfect negative correlation

$r = 1$ means perfect positive correlation

$r = 0$ means zero correlation

$r = -0.84$ means strong negative correlation

$r = 0.15$ means weak positive correlation

and so on.

A simple worked example for calculating r

Let us suppose that you have a simple data set consisting of only the following three pairs of values

X	Y
1	2
2	3
3	4

Step 1 Calculating the covariance, S_{XY}

X	Y	XY
1	2	2
2	3	6
3	4	12
$\Sigma X = 6$	$\Sigma Y = 9$	$\Sigma XY = 20$

$$\overline{X} = \frac{6}{3} = 2 \quad \overline{Y} = \frac{9}{3} = 3$$

$$S_{XY} = \frac{20}{3} - 2 \times 3$$

$$= 6\frac{2}{3} - 6 = \frac{2}{3}$$

Step 2 Calculating the standard deviation of X, S_X

X	X^2
1	1
2	4
3	9
$\Sigma X = 6$	$\Sigma X^2 = 14$

$$\overline{X} = 2$$

$$S_X = \sqrt{\frac{14}{3} - (2)^2}$$

$$= \sqrt{\frac{2}{3}}$$

Step 3 Calculating the standard deviation of Y, S_Y

Y	Y^2
2	4
3	9
4	16
$\Sigma Y = 9$	$\Sigma Y^2 = 29$

$$\bar{Y} = 3$$

$$S_Y = \sqrt{\frac{29}{3} - (3)^2}$$

$$= \sqrt{\frac{2}{3}}$$

Finally, $r = \dfrac{S_{XY}}{S_X S_Y} = \dfrac{\frac{2}{3}}{\sqrt{\frac{2}{3}} \times \sqrt{\frac{2}{3}}} = \dfrac{\frac{2}{3}}{\frac{2}{3}} = 1$

So in this example there is perfect positive correlation. When the three points are plotted on a scattergraph, this should make sense just by looking at their pattern. As you can see from Figure 8.6, the points form a perfect straight line with a positive slope.

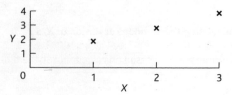

Figure 8.6 *Data from simple worked example showing perfect positive correlation.*

There are two important provisos which need to be considered when interpreting correlation.

The first point is the question of *batch size*. In general, the smaller the batch size, the larger must be the value of *r* in order to provide evidence of significant correlation.

Secondly, a crucial issue here is to be able to distinguish correlation between two variables and a *cause-and-effect relationship* between them. It is important to remember that correlation only suggests a pattern linking two sets of data. CORRELATION DOES NOT PROVE THAT ONE OF THE VARIABLES HAS *CAUSED* THE OTHER. Indeed, exactly what the value of *r* does and does not tell us is quite difficult to interpret, and this issue is pursued in more detail in the final section of this chapter.

Rank correlation

It is not always necessary, or even possible, when investigating correlation, to draw on the sort of measured data that have been presented so far. An alternative is to work only from the rank positions. For example, when judging ice-skating or gymnastics competitions, the marks that are awarded are rather meaningless and their primary purpose is to allow the competitors to be ranked in order – first, second, third, and so on. Information is sometimes stripped of the original data and the results presented only in order of rankings.

While it would be possible to use the correlation coefficient, r, in such cases, there is an alternative measure which is specially designed for ranked data called the rank coefficient of correlation, r'. It is sometimes known as 'Spearman's coefficient of rank correlation' after its inventor, Charles Spearman.

The formula for r' is based on calculating all the differences (d) between each pair of ranks.

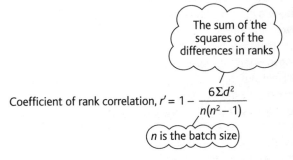

The sum of the squares of the differences in ranks

Coefficient of rank correlation, $r' = 1 - \dfrac{6\Sigma d^2}{n(n^2 - 1)}$

n is the batch size

As with the product-moment correlation coefficient, r, the coefficient of rank correlation, r', has been devised to ensure that its value will lie within the range -1 to 1. The value for r' can be interpreted in much the same way as the values for r were in the previous section. Thus:

$r' = -1$ means perfect negative correlation
$r' = 1$ means perfect positive correlation

$r' = 0$ means zero correlation
$r' = -0.84$ means strong negative correlation
$r' = 0.15$ means weak positive correlation

and so on.

Cause and effect

For certain everyday events, the link between cause and effect seems fairly straightforward.

However, a strong correlation between two things does not prove that one has *caused* the other. A strong correlation merely indicates a statistical link, but there may be many reasons for this besides a cause-and-effect relationship.

For example, traditionally the height of a pregnant woman and her shoe size were thought to be good predictors of whether or not she would need to give birth by Caesarean section operation. In general, it was thought, women who were short in stature and petite of foot would also have narrow pelvises and would therefore find it difficult to have a normal vaginal delivery. And indeed, the statistical evidence seems to bear this out, in that there has traditionally been a close correlation between women having a Caesarean birth and their having small feet. But this is not the same thing as saying that there is a close correlation between women needing a Caesarean birth and having small feet. It now seems that this is something of a myth which has grown up over time. Doctors, believing the association to be valid, tend to prescribe Caesarean births for women with small feet, thereby making the correlation a self-fulfilling prophecy.

An association between two things is known as *spurious correlation* when it implies a cause-and-effect link which is not actually there.

We really can't be sure which one was the cause and which the effect, or indeed whether both things were quite independent of each other and the changes were because of some other factor or factors.

Cause and effect can really only be tested under controlled conditions where all other possible influences are removed. Then one factor (the independent one) is systematically altered and the resulting change on the other one (the dependent factor) is observed.

9

chance
and
probability

Probability is a way of describing the likelihood of events taking place. Clearly different events have different likelihoods of occurring. At one extreme, some events may be thought of as being *certain* (the chance that you will eventually die, for example,) while at the other extreme another event might be described as *impossible* ('flying pigs', maybe). But for most everyday situations, the degree of likelihood involved will lie somewhere between these two extremes.

The language of chance employs terms such as 'very likely', 'an even chance', 'a long shot', and so on. However, there are other situations where words alone are just not sufficiently precise to describe and compare probabilities accurately. In the world of betting, probabilities are measured using odds. In statistics, probabilities are measured using a number scale in the range 0 to 1.

Two key ideas in probability are mutually exclusive events and the notion of independence.

Odds

Odds are the most common way of measuring uncertainty in situations where people are betting on an event whose outcome is unknown. In a horse race, for instance, the odds on each horse are a measure of how likely the bookmaker thinks each horse is to win a particular race. For example, if a horse is given odds of 5–1 (said as 'five to one' and usually written either as 5–1 or 5:1) this means that, given six chances, it would be expected to win once and lose the other five times. Suppose you bet on a horse with odds of 10:1 and the horse wins, then for each £1 you bet, you win £10. If, instead, you had placed a bet of £5 you would win £50 as well as receiving the £5 that you had staked in the first place.

If a horse is favourite to win, it is given short odds, perhaps 2:1 or 3:2; whereas unlikely winners will be given very long odds, perhaps 100:1. A near certainty might be described in betting jargon as an 'odds-on favourite'. This is where a horse is given such short odds that a winning bet will earn the punter less than their stake. Of course, they will get their stake back also. An example of an odds-on favourite might be odds of '2 to 1 on'. These are actually odds of '1 to 2'. In other words, betting £2 wins you a mere £1.

Statistical probability

In statistical work, probabilities are usually measured as numbers between zero and 1 and can be expressed either as fractions or as decimals. A probability of zero ($p = 0$) means zero likelihood, i.e. the outcome is impossible. A probability of one ($p = 1$) refers to an outcome of certainty. Usually, in statistical work, outcomes are the results of a *trial* or *experiment* (like 'tossing a coin', for example). Most trials have outcomes whose probabilities lie somewhere between zero and one. Thus, an outcome thought of as being fairly unlikely might have a probability of something like 0.1, while an 'even chance' would correspond to a probability of 0.5, and so on.

Myths and misconceptions about probability

There are many examples of faulty intuitions in the area of probability. Indeed it is quite possible that you have some yourself. Exercise 9.1 will give you an opportunity to test out your intuitions in this area.

Exercise 9.1 Beliefs and intuitions

Which of the following do you think are true?

i) If you throw a six-faced die, a 'six' is the most difficult outcome to get.

ii) If a coin is thrown six times, three heads and three tails is the most likely outcome.

iii) If a coin is thrown six times, of the two possible outcomes below, the second is more likely than the first (H = Heads, T = Tails)

Outcome A   H H H H H H
Outcome B   H T T H H T

iv) If two dice are thrown and the results added together, scores of 12 and 7 are equally likely.

v) If a fair coin was tossed ten times and each time it showed 'heads' you would expect the eleventh toss to show 'tails'.

[Comments below]

(i) Although many people believe that a six is the most difficult score to get when a die is tossed, this is actually false. There are several possible explanations for why this mistaken notion might persist. One may be that six is the biggest score, and therefore thought to be hardest to throw. A second may be that in many dice games, six is the score that is required and therefore it seems logical that it should be the most difficult to get. But perhaps the most likely explanation is that people correctly feel that it is less likely to toss a six than to toss a 'not-six', and therefore it is more difficult to get a six, than not to get a six. However, this is quite a different belief from that of thinking that the six outcomes on a die are not equally likely.

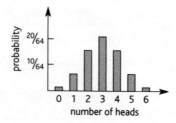

Figure 9.1 *Bar chart of the likelihoods of the possible outcomes if a coin is thrown six times.*

(ii) This statement is correct. There are seven possible outcomes for this event, which range from getting no heads, to getting six heads. The two least likely outcomes are the extremes of 'all six heads' and 'all six tails', whereas three heads and three tails is the most likely. The bar graph, Figure 9.1, summarizes the likelihoods of the various outcomes. Just why the graph looks like this and how to calculate the separate probabilities of each outcome (indicated on the vertical scale) are not covered here but are to do with something called 'The binomial distribution'.

(iii) This statement is false. Although an outcome with three heads and three tails is more likely than an outcome with six heads, the two sequences given here are in fact equally likely. The reason for this is that there is only one way of getting six heads in a row (HHHHHH), whereas there are many ways of getting three heads and three tails – for example, HTHTHT or HHHTTT or TTTHHH, etc. Taking order into account, all of these sequences are equally likely, so HHHHHH is as likely as HTTHHT.

(iv) This statement is false. There is only one way to throw a '12' (with a double six), whereas there are many ways in which a 'seven' can be obtained. The bar graph, Figure 9.2, shows the likelihood of each of the 11 outcomes (2, 3, ... 12) when two dice are thrown.

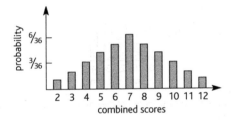

Figure 9.2 *Bar chart of the likelihoods of the possible outcomes if two dice are thrown.*

(v) The final statement is also false. Many people believe that, according to some vague sense of the ill-defined 'law of averages', an unexpectedly long run of heads will necessarily be corrected for by a tail on the next throw. Whether it is the coin itself or God who is responsible for this even-handed averaging out of outcomes is unclear. The reality is that the coin has no memory of past outcomes and begins each new toss afresh. This misconception is sometimes known as the 'gambler's fallacy' as it leads punters to believe that, after a string of losses, their next bet is more likely to win.

Mutually exclusive outcomes and adding probabilities

When two (or more) outcomes are said to be 'mutually exclusive' this means that if one occurs, the other cannot occur. For example, if a coin is tossed, the two possible outcomes 'getting heads' and 'getting tails' are mutually exclusive, because both cannot occur at the same time. Not all trials produce outcomes which are mutually exclusive. For example, when a baby is born, one outcome could be that it is a girl (rather than a boy). Another outcome might be that it has brown eyes (rather than blue or green). However, if a brown-eyed girl is born, both outcomes have occurred at the same time – clearly not mutually exclusive outcomes.

In general, if a trial has several possible mutually exclusive outcomes – A, B, C, etc. – the probabilities are written as follows:

> The probability of outcome A resulting is $P(A)$
> The probability of outcome B resulting is $P(B)$

and so on.

If we want to find the probability of *either* A *or* B occurring, we *add* separate probabilities for A and B.

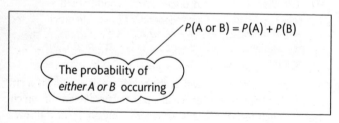

$$P(A \text{ or } B) = P(A) + P(B)$$

The probability of *either* A *or* B occurring

However, this rule only applies if outcomes A and B are mutually exclusive, otherwise the area of intersection has been double counted.

If two outcomes are not mutually exclusive, the rule must be adjusted to take account of the double counting, as follows.

For events that are not mutually exclusive:

$$P(A \text{ or } B) = P(A) + P(B) - P(A \text{ and } B)$$

The probability of *either A or B* occurring

The probability of *both A and B* occurring

Independence and multiplying probabilities

Exercise 9.2 Tossing and turning

Suppose a fair coin was tossed ten times and each time it showed 'heads'. What would you expect the eleventh toss to show?

[Comments below]

Notice that the question states at the outset that the coin is 'fair'. This means that you can assume there is no bias either towards 'heads' or 'tails'. Also, under normal conditions, the successive tosses of a coin are independent events. In other words, the outcomes of the first ten tosses will not influence the outcome of the eleventh toss. Therefore, given that the coin is fair, the chance of heads or tails on the eleventh toss is still 50–50. Of course, if you didn't know that the coin was fair you might wish to argue that there was evidence of a bias towards 'heads', in which case you might feel that 'heads' was more likely on the next throw. However, there is never any argument for thinking that 'tails' is more likely after a run of heads and, interestingly, this is the most commonly held view.

Exercise 9.3 Lucky dip

A lucky dip consists of 20 envelopes, only 3 of which contain a prize.

a) On the first dip, what is the probability of winning a prize?

b) Suppose the first envelope chosen did contain a prize and the envelope is removed from the bag. What is the probability of the second dip also winning a prize?

[Comments below]

(a) Using the definition of probability given earlier, the probability of winning on the first dip

$$= \frac{number\ of\ prizes}{number\ of\ possibilities} = \frac{3}{20}$$

(b) After the first dip, the conditions for the second dip have been altered. In the latter case, the required probability

$$= \frac{2}{19}$$

In this instance, the second event (dip number 2) is *dependent* on the first (dip number 1).

Exercise 9.4 Three in a row!

(a) Suppose you toss a fair coin three times in succession. What is the probability of all three coming up 'heads'?

(b) Suppose that you were able to take three dips in the lucky dip described in Exercise 9.3. What is the probability that you will win each time?

[Comments below]

(a) Since successive tosses of a fair coin are independent events, it is perfectly legitimate to multiply each of the separate probabilities involved. Thus:
Probability of three 'heads' in a row = (1/2) × (1/2) × (1/2) = 1/8

(b) As was explained in the comments to Exercise 9.3, the probability of being successful in the second and third dips has altered as a result of the previous outcomes. Thus: Probability of winning a prize in all three dips

$$= 3/20 \times 2/19 \times 1/18$$
$$= 6/6840 \text{ (or 1 chance in 1140)}$$

In summary, then, if two trials are independent, the outcome of one does not affect the outcome of the other.

10

deciding on differences

This chapter introduces a major area of statistics called tests of significance, otherwise known as hypothesis testing. A key idea here is that, due to natural variation, you would expect some degree of difference to occur just by chance. The essential principle underlying significance testing is that the difference being observed must be sufficiently large to warrant reaching the conclusion that there is a 'real' difference, and unlikely to be one due to chance alone. Two types of error are considered – known as Type 1 and Type 2 error. A Type 1 error occurs where a difference is claimed when none actually exists. A Type 2 error is the converse of this – i.e. where there really is a difference between the things being tested but the test fails to pick this up. The chapter ends with a brief look at the four main stages involved in carrying out a test of significance.

As the title suggests, the central theme of this chapter is about how we can decide whether or not observed differences can be described as being 'statistically significant' (in other words, indicate some real underlying difference) or simply the result of chance variation.

Contexts for investigating differences

A quick glance through any daily newspaper will produce a variety of situations where the process of investigating differences has occurred. Table 10.1 shows a few examples.

Table 10.1 *Examples of 'differences' that crop up in newspapers*

Headline	Basic story
Top doctors condemn alternative medicine	Members of the Royal College of Physicians claim that 'pseudo-scientific' alternative medicine is often an irrational waste of time and money and can even be a serious risk to health. 'People have been misled by the placebo effect, which comes from suggestion, not medical intervention', they claimed.
Child smokers are more likely to take drugs, survey finds	According to a Health Education Authority survey, more than half the children who smoke regularly have been offered drugs and half have tried them. They are also more likely to drink alcohol. The survey, conducted by MORI, was based on interviews with 10 000 nine to 15-year-olds.
Most heart disease in North	According to a Health Education Authority report, a survey into coronary heart disease has confirmed that death rates are highest in the Northern, Yorkshire and Northwestern regions and lowest in the Thames, Wessex, Oxford and East Anglia regions.

The common theme in these three health surveys, and in many other newspaper reports like them, is that researchers have made some sort of comparison between two or more things and 'proved' that there is a 'significant difference' between them – i.e. a difference that could not merely be explained by natural variation in the data. The notion of what constitutes 'proof' is worth examining more closely here. A statistician's view of 'proof' is rather different to that of the mathematician. Whereas a mathematical result is normally 'proved' with 100% certainty, a statistical proof is closer to the legal notion of a conclusion being proved 'beyond reasonable doubt'. The difference is that mathematicians are able to create inside their heads a perfect world of numbers, symbols and relationships in which absolute truth can exist. Statisticians, like juries, have to take the real world as they find it and make the best decisions they can under conditions of uncertainty. All statistical judgements, therefore, are given in association with a measure of probability attached to them, which indicates how confident we might be that the pattern is a real one and not just some fluke result due to chance. However, we can never rule out the possibility that it just might be a fluke result.

Some basic terminology

And now the terms 'tail' and 'critical value' require explanation.

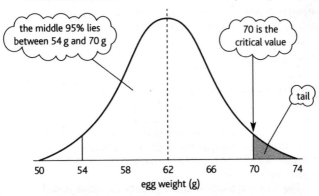

Figure 10.1 *Some features of a distribution.*

Term	Explanation
Tail	The tail is the small area to the left and to the right of the distribution. The tail can be thought of either as 'the proportion of items falling into this extreme range' or alternatively, 'the probability of a randomly chosen value from the distribution falling into this extreme range'. Both of these definitions basically amount to the same thing. The tail shaded in Figure 10.1 is a measure of the probability that a hen egg, chosen at random, would weigh over 70 g, and this works out at roughly $2\frac{1}{2}$%.
Critical value	The critical value of a distribution of values is a boundary value which marks the divide between the main body of the distribution and the tail. If the value being tested lies outside the critical value (i.e. if it falls into the tail), it is deemed to be 'significantly different' from the mean of the distribution.

Some additional terminology is needed in the discussion of significance testing and the next collection of terms is introduced by means of the legal world of the courtroom.

It has already been suggested in this chapter that there are parallels between the ways in which decisions are made in a court of law and in statistics. In this section, the courtroom metaphor is extended and developed in order to illustrate some key terms and ideas in statistical decision-making.

Below is a fairly typical summing-up speech which a judge might conceivably make to a jury shortly before they retire to make their decision on the innocence or guilt of the accused. The underlined words parallel certain important statistical ideas which are explained underneath the extract.

Members of the jury, you have heard the evidence. I would like to remind you that in our system of justice, an accused person is innocent until proved guilty and so the burden of proof lies with the prosecution, not the defence. I urge you, therefore, to <u>start out with the assumption of innocence</u>.

Remember that absolute 100 per cent certainty is never possible to achieve, but you need to <u>be convinced beyond all reasonable doubt</u> of the guilt of the accused before delivering a 'guilty' verdict.

You must bear in mind the consequences of a wrong decision. <u>If you find the accused 'guilty' when he is actually innocent</u>, then an innocent person is sent to jail. But <u>if you find the accused 'innocent' when he is actually guilty</u>, then a villain goes free. It is the view of this court that <u>better a thousand guilty villains go free than one innocent person is wrongly convicted</u>.

There now follows a 'translation' of these five courtroom phrases from 'legal-speak' into 'statistics-speak' in the statistical area that is the subject of this chapter — testing for whether there is a significant difference between two things.

(a) Starting out with the assumption of innocence: When testing for a difference in statistics, we start by assuming that there is no difference between the values being tested. This is stated as a hypothesis of no difference and is called the null hypothesis (written as H_0). The 'alternative hypothesis' (written as H_1) is the alternative explanation, namely that there is a real difference between the values being tested. It is only when the null hypothesis is rejected that the 'alternative hypothesis' can be accepted.

(b) Being convinced beyond all reasonable doubt: As the judge remarked, in the real world, 'absolute 100 per cent certainty is never possible to achieve'. In statistics we talk about differences being 'significant to, say, 0.05 (i.e. 5%) or, perhaps, 0.01 (i.e. 1%)'. These figures correspond to the levels of 'reasonable doubt' that the differences might have been just a fluke. The smaller the level of significance, the smaller the 'reasonable doubt' that the differences were due to a fluke result. Thus, a test of significance at the 1% level of significance is a more rigorous test than one at the 5% level of significance.

(c) Finding the accused 'guilty' when he is actually innocent: This sort of error is known in statistics as a 'Type 1' error and involves believing that there is a difference between the values when in fact there is no difference. In other words, a Type 1 error is where you accept H_1 when you should have accepted H_0.

(d) Finding the accused 'innocent' when he is actually guilty: As you might expect from reading (c) above, this is known in statistics as a 'Type 2' error and it occurs where you believe there is no difference between the values when in fact there is a difference – i.e. you accept H_0 when you should have accepted H_1.

(e) Better a thousand guilty villains go free than one innocent person is wrongly convicted: This statement says something about the relative importance of the two types of error described above. If, in a statistical test of significance, you apply a very stringent test (say, applying a low level of significance of 0.001, or 0.1%) this means that you will only accept H_1 if there is an extremely large difference between the values. However, such circumstances make it likely that there may be 'real' differences which are not picked up – in other words, this is a situation where a Type 2 error is likely. However, the other side of the coin is that, under these conditions of a low significance level, you are very unlikely to claim a difference when none exists – i.e. you are unlikely to make a Type 1 error. Correspondingly, in a test of significance with a high level of significance (say 0.1 or 10%), you are likely to make a Type 1 error but unlikely to make a Type 2 error. The message from the judge to statisticians seems to be – 'Set your significance levels low to ensure that you are much more likely to commit a Type 2 error than a Type 1 error. This means that you will miss spotting some significant differences but the differences that you do claim are very likely to be true.'

You may have found some of the terminology and explanations above rather difficult to follow, so don't be afraid to read these five points, (a) to (e) more than once. Table 10.2 summarizes the essential relationship between the two types of error, both in the context of a court-room and in statistical decision-making.

Table 10.2 *Type 1 and type 2 errors*

(a) *In a court room*

		Verdict	
		Not-guilty	*Guilty*
Defendant	*Saint*	A correct decision	Wrongly accused
	Villain	Wrongly acquitted	A correct decision

(b) *In statistical decision-making*

		Conclusion	
		No difference (H_0)	*Difference (H_1)*
Reality	Polulations really are the same	A correct decision	Type 1 error
	Populations really are different	Type 2 error	A correct decision

What is a test of significance?

The terms 'test of significance' and 'hypothesis test' both refer to more or less the same statistical process, of deciding whether or not there is a difference between two values or two sets of results. There are a variety of tests to choose from, the particular choice depending on the nature of the data and the context in which the comparison is being made. However, in general, the procedure for carrying out a test of significance can usually be summarized into four clear stages. These are set out in Figure 10.2.

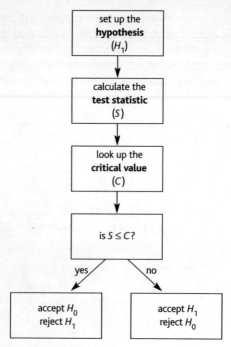

Figure 10.2 *Flow chart showing the four main stages of a test of significance.*

Notes

Notes